Collins
INTERNATIONAL

C000247014

Science
Foundation
Anthology

Published by Collins
An imprint of HarperCollins*Publishers*
The News Building, 1 London Bridge Street,
London, SE1 9GF, UK

HarperCollins*Publishers*
Macken House, 39/40 Mayor Street Upper,
Dublin 1, DO1 C9W8, Ireland

Browse the complete Collins catalogue at
www.collins.co.uk

ISBN 978-0-00-846893-4

British Library Cataloguing-in-Publication Data
A catalogue record for this publication is available from the British Library.

Compiled by: Fiona Macgregor
Publisher: Elaine Higgleton
Product manager: Letitia Luff
Commissioning editor: Rachel Houghton
Edited by: Eleanor Barber
Editorial management: Oriel Square
Cover designer: Kevin Robbins
Cover illustrations: Jouve India Pvt Ltd.
Internal illustrations: p 4cr Sylwia Filipczak, p 8–9 Laszlo Veres,
p 13 Beccy Blake, p 14–15 Camilla Galindo, p 26–27 Jon Stuart,
p 30, 32 Tasneem Amiruddin
Typesetter: Jouve India Pvt. Ltd.
Production controller: Lyndsey Rogers
Printed and Bound in the UK using 100% Renewable
Electricity at Martins the Printers

Acknowledgements

With thanks to all the kindergarten staff and their schools around the world who have helped with the development of this course, by sharing insights and commenting on and testing sample materials:

Calcutta International School: Sharmila Majumdar, Mrs Pratima Nayar, Preeti Roychoudhury, Tinku Yadav, Lakshmi Khanna, Mousumi Guha, Radhika Dhanuka, Archana Tiwari, Urmita Das; Gateway College (Sri Lanka): Kousala Benedict; Hawar International School: Kareen Barakat, Shahla Mohammed, Jennah Hussain; Manthan International School: Shalini Reddy; Monterey Pre-Primary: Adina Oram; Prometheus School: Aneesha Sahni, Deepa Nanda; Pragyanam School: Monika Sachdev; Rosary Sisters High School: Samar Sabat, Sireen Freij, Hiba Mousa; Solitaire Global School: Devi Nimmagadda; United Charter Schools (UCS): Tabassum Murtaza; Vietnam Australia International School: Holly Simpson

The publishers wish to thank the following for permission to reproduce photographs.

(t = top, c = centre, b = bottom, r = right, l = left)

p 2b, p 3t, p 12 Will Amlot, p 23t Jochen Tack/Alamy Stock Photo, p 23b Rawpixel Ltd/Alamy Stock Photo, p.24–25 Steve Lumb, p 28–29 Mark Coote. All other photographs: Shutterstock.

MIX
Paper | Supporting
responsible forestry
FSC™ C007454

I am a scientist

I look.

I measure.

I try new things.

I say what I know.

Living and non-living

Which things are living?

Which things are non-living?

I am a living thing.

I breathe and I move.

I eat and I grow.

My senses

eyes

ears

nose

mouth

Which ones feel hot?

Which ones feel cold?

Remember, don't touch hot things!

Animals

Can you find the animals?

What other animals can you find?

Where do these animals live?

Which animals are the same?

How are the animals different?

Plants

Different plants have different leaves.

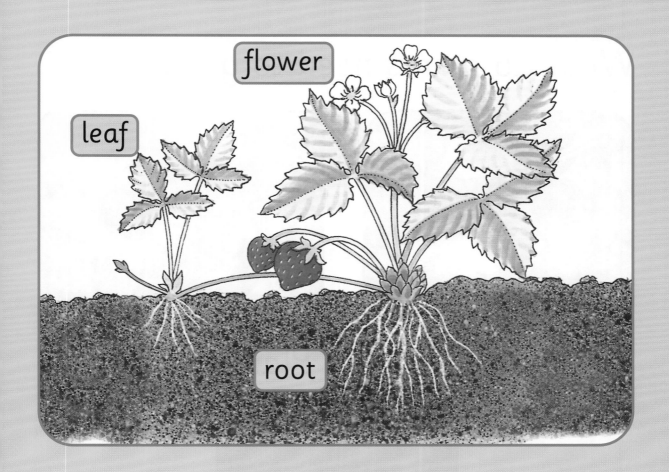

flower

leaf

root

What plants and animals need

What do plants need to grow?

Plants need soil.

Plants need sun.

Plants need water.

Animals need food, water and shelter.

Habitats

A lot of birds live in trees.

Some birds build nests.

Is this a bird's nest?

No, it's a beehive.

A fox lives in a den.

A bat lives in a cave.

Water

We drink water.

We use water every day.

Growing and changing

Animals have babies.

A cat and
its kittens.

Two hens and
their chicks.

A sheep and
its lambs.

A goat and its kid.

Wild animals have babies too.

Two lionesses and their cubs.

A grizzly bear and its cubs.

Four monkeys and their infants.

A duck and its ducklings.

Light and dark

It is day.

It is night.

oil lamp

candle

electric light

the sun

Weather

It is raining.

I wear a raincoat and boots.

It is hot. I put on sunscreen.

We have a picnic.

Materials

How to make a sock puppet.

1

Tuck the sock in.

2

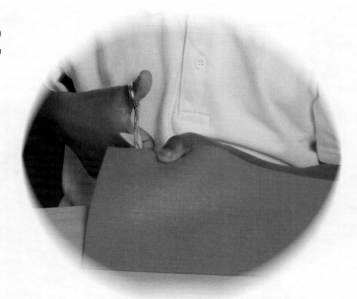

Cut a tongue out.

3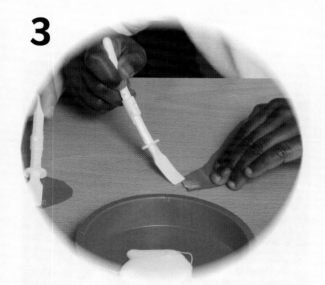

Stick the
tongue on.

4

Cut and stick
some eyes on.

5

Enjoy using your puppet!

Pick up the litter.

Put it in the bag.

Pick up a can.

Run and tip it in.

Pushes and pulls

Pushing

Pulling

Changing materials

Things I can do.

stretch

bend

twist

squash

Does it stretch?

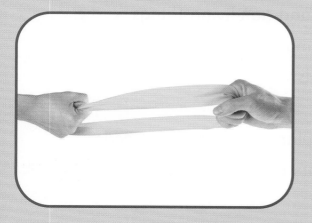

Does it bend?

Does it twist?

Does it squash?

I know about:

- [ ] being a scientist
- [ ] living and non-living things
- [ ] using my senses
- [ ] animals
- [ ] plants
- [ ] what plants and animals need
- [ ] habitats of animals
- [ ] water
- [ ] growing and changing
- [ ] light and dark
- [ ] the weather
- [ ] materials
- [ ] taking care of our world
- [ ] pushes and pulls
- [ ] changing materials.